THAT'S THE JOB FOR ME!

I'M GOING TO BE AN ENGINEER!

BY MICHOU FRANCO

Gareth Stevens
PUBLISHING

Please visit our website, www.garethstevens.com. For a free color catalog of all our high-quality books, call toll free 1-800-542-2595 or fax 1-877-542-2596.

Library of Congress Cataloging-in-Publication Data

Names: Franco, Michou, author.
Title: I'm going to be an engineer! / Michou Franco.
Description: Buffalo, NY : Gareth Stevens Publishing, [2025] | Series: That's the job for me! | Includes index.
Identifiers: LCCN 2023039979 (print) | LCCN 2023039980 (ebook) | ISBN 9781538293416 (library binding) | ISBN 9781538293409 (paperback) | ISBN 9781538293423 (ebook)
Subjects: LCSH: Engineering–Vocational guidance–Juvenile literature. | Engineers–Juvenile literature.
Classification: LCC TA157 .F7155 2025 (print) | LCC TA157 (ebook) | DDC 620.0023–dc23/eng/20230912
LC record available at https://lccn.loc.gov/2023039979
LC ebook record available at https://lccn.loc.gov/2023039980

Published in 2025 by
Gareth Stevens Publishing
2544 Clinton Street
Buffalo, NY 14224

Designer: Claire Zimmermann
Editor: Therese Shea

Photo credits: Cover, p. 1 Nopphon_1987/Shutterstock.com; Series Art (background) Salmanalfa/Shutterstock.com; p. 5 Amorn Suriyan/Shutterstock.com; p. 7 Chaosamran_Studio/Shutterstock.com; pp. 9, 17 Gorodenkoff/Shutterstock.com; p. 11 EF Stock/Shutterstock.com; p. 13 Narin Nonthamand/Shutterstock.com; p. 15 DC Studio/Shutterstock.com; p. 19 BELL KA PANG/Shutterstock.com; p. 21 Sorapop Udomsri/Shutterstock.com.

Printed in the United States of America

CPSIA compliance information: Batch #CS25GS: For further information contact Gareth Stevens, New York, New York at 1-800-542-2595.

CONTENTS

Boldface words appear in the glossary.

I'm Into Engineering

I love to build things. I love to take things apart and put them back together. I love to work with a team. I *really* love to **solve** problems. Math problems are my favorite! So, I'm going to be an engineer.

What Is an Engineer?

Engineers use math and science to make our world better. They may build something, such a machine. They may change a **structure** to make it better, such as a bridge. They may also **design** a process, or way of making or doing something.

Kinds of Engineers

What kind of engineer will I be? There are many different kinds! A civil engineer designs and builds structures. Electrical engineers work with **electricity**. Mechanical engineers work with machinery. Aerospace engineers send things into space! Other engineers work with computers or **chemicals**.

Mechanical Engineering 101
Simplified

Engineering Students

All engineers go to college. That's a school after high school. They take classes that prepare them for a job. They learn different kinds of math. They also take science classes such as physics. Physics is the study of matter, **energy**, force, and movement.

Some engineers go to school after college. They may study more about a certain kind of engineering. They may learn how to be a manager, or leader on the job. But other engineers start jobs right away. Some need a special **license**. They pass a test to get it.

On the Job

All engineers follow steps to make sure their **projects** will work. First, they design. They use their knowledge to come up with the best plan they can think of. They often use special computer **software** to do this.

ArchiPlanner 28.11f
Version Control
Today
Thursday
Wednesday
Tuesday
Monday
Layout 1
Layout 2

The engineer uses their design to build a prototype. A prototype is a model, or an example. They can test their prototype to find any problems with the design. Then, they fix the problems before they start on the real thing!

Teamwork

Engineers often work as a team. They show their designs and prototypes to other engineers. They listen to each other's ideas. For large jobs, engineers may each have a task to do. Together, they can solve problems that come up.

Getting Ready

I can get ready for my engineering job now. I can try hard in my math and science classes. I can do projects that teach me engineering skills. I'm building a robot right now. I can't wait to be an engineer!

GLOSSARY

chemical: Matter that can be mixed with other matter to cause changes.

design: To create the pattern or shape of something.

electricity: The flow of electrical power, which is the movement of charged particles, usually electrons, between atoms.

energy: Power used to do work.

license: A piece of paper that allows someone to do something.

project: A planned piece of work that has a certain goal.

software: A computer code, or program, that tells a computer what to do.

solve: To find the answer.

structure: Something that is built.

FOR MORE INFORMATION

BOOKS

Jacoby, Jenny. *Civil Engineering: Engineering Is All Around You.* New York, NY: Kingfisher, 2022.

Schuette, Sarah L. *10-Minute Engineering Projects.* North Mankato, MN: Capstone Press, 2020.

Smibert, Angie. *Engineering*. Mankato, MN: The Child's World, 2021.

WEBSITES

Engineering Quiz
www.sciencekids.co.nz/quizzes/engineering.html
Do you know a bit about famous structures? Test your knowledge.

NASA for Kids: Intro to Engineering
education.nationalgeographic.org/resource/nasa-kids-intro-engineering/
The U.S. space agency talks about the engineering process on this site.

Publisher's note to educators and parents: Our editors have carefully reviewed these websites to ensure that they are suitable for students. Many websites change frequently, however, and we cannot guarantee that a site's future contents will continue to meet our high standards of quality and educational value. Be advised that students should be closely supervised whenever they access the internet.

INDEX